Lunar Orbital Platform-Gateway

by Patrick H. Stakem

2017

Number 16 in the Space series.

Table of Contents

Introduction..3
Author...3
Predecessors...4
   Skylab...4
   Skylab-II..5
   Salyut..5
   MIR..6
   Shuttle-Mir Program..7
   ISS..8
   Decommissioning of the ISS...11
   Lessons Learned from similar Earth-based facilities.......12
   Deep Space Gateway 13
LOP-G...16
Follow-ons..16
   Deep Space Habitat...16
   Russian Lunar Orbital Station 17
   NASA Lunar Outpost..18
Transportation and Logistics..19
Mars Base Camp..20
Broader Space Colonization...23
Wrap-up..24
Bibliography...26
Resources..31
Glossary of terms and definitions..33
If you enjoyed this book, you might also be interested in some of these..38

"The goal isn't just scientific exploration.... It's also about extending the range of human habitat out from Earth into the solar system as we go forward in time.... In the long run a single-planet species will not survive.... If we humans want to survive for hundreds of thousands or millions of years, we must ultimately populate other planets."

Michael D. Griffin, former NASA Administrator

Introduction

This book covers the topic of the Lunar Orbital Platform-Gateway (LOP-G), a renaming and restructuring of the Deep Space Gateway, a joint Russian-US effort, and associated missions. This is a step beyond the International Space Station, which will be beyond its useful lifetime in a few years, and will be decommissioned, with some parts being reused, and some re-entered. This will result in a new era of human space exploration, further from Earth. Whether we refer to the emerging facility as a Gateway, a Colony, a settlement, or a habitat, we are talking of a permanently occupied facility. We can consider the habitat to be in orbit (about something), or on the surface of another body, other than Earth. These projects will differ in detail, but will all consist of self-sufficient structures somewhere other than Earth, with an associated logistics train. The Gateway would be continuously crewed, and serves as an outpost form which to explore the lunar and Martian surfaces.

Author

Mr. Patrick H. Stakem has been fascinated by the space program since the Vanguard launches in 1957. He received a Bachelors degree in Electrical Engineering from Carnegie-Mellon University, and Masters Degrees in Physics and Computer Science from the Johns Hopkins University. At Carnegie, he worked with a group of undergraduate students to re-assemble, modify, and operate a surplus missile guidance computer, which was later donated to the

Smithsonian. He was brought up in the mainframe era, and was taught to never trust a computer you could lift.

He began his career in Aerospace with Fairchild Industries on the ATS-6 (Applications Technology Satellite-6) program, a communication satellite that developed much of the technology for the TDRSS (Tracking and Data Relay Satellite System). He followed the ATS-6 Program through its operational phase, and worked on other projects at NASA's Goddard Space Flight Center including the Hubble Space Telescope, the International Ultraviolet Explorer (IUE), the Solar Maximum Mission (SMM), some of the Landsat missions, and Shuttle. He was posted to NASA's Jet Propulsion Laboratory for Mars-Jupiter-Saturn (MJS-77), which later became the *Voyager* mission, and is still operating and returning data from outside the solar system at this writing. He initiated and lead the international Flight Linux Project for NASA's Earth Sciences Technology Office. He is the recipient of the Shuttle Program Manager's Commendation Award, and has completed 42 NASA Certification courses. He has two NASA Group Achievement Awards, and the Apollo-Soyuz Test Program Award.

Mr. Stakem has been affiliated with the Whiting School of Engineering of the Johns Hopkins University since 2007, and Capitol Technology University. Mr. Stakem supported the Summer Engineering Bootcamp Projects at Goddard Space Flight Center for 2 years

Predecessors

There have been a series of Space Stations in Earth orbit, the most recent being the International Space Station. The ISS was built with lessons learned from the Russian MIR Station, and the Shuttle-MIR program. Earlier missions are discussed below.

Skylab

Skylab was an Apollo-era U.S. Space Station. It used a Saturn S-IVB upper stage as the structure for the station, launched by a Saturn-V with live first and second stages. The hydrogen fuel tank was re-purposed into the manned facility. The oxygen tank was used for trash and disposables. The payload to orbit was 170,000 pounds. The station was 82 feet long, 56 feet wide, and 36 feet in the other direction. It was quite visible from Earth. It was serviced by Apollo capsules for both logistics, crew delivery, and auxiliary electric power. This was pre-Shuttle, and even after Skylab was deliberately re-entered for safety reasons, another Skylab could have been launched and then serviced by the Shuttle.

Skylab-II

Skylab-II was a circa 2013 concept from the Marshall Space Flight Center's Advanced Concepts Office. It would be the same concept as the original Skylab, but using the upper stage hydrogen fuel tank from the Space Launch System then under development. It was to be located at the Earth-Moon L2 point (Lagrange point, a null in the gravity field). Here, it would need minimal orbital adjust to remain at that spot. That particular point is on the other side of the moon, from the Earth. That puts it some 430,000 km from Earth, and 62,800 km from the lunar surface. With the moon between the station and the Earth, a relatively quiet radio environment is achieved. The goal is to support a 4 person crew for 60 days, without a resupply flight. The re-purposed tank would have a diameter of 8.5 meters, larger than the ISS's 4.5 meters. The provided volume would be about 500 cubic meters. Lessons learned from the ongoing ISS mission would be applied to the Skylab-II project.

Salyut

Russia's on-orbit experience with Space Stations started with the Salyut in 1971. On the first crewed flight to Salyut-1, the crew was unable to dock, due to a failure in the mechanism on their craft. A second mission was successful, and the crew occupied the facility for more than 3 weeks. Unfortunately, the crew as killed while returning, due to a faulty pressurization valve. They were not wearing pressure suits, as the cabin space was not sufficient. Salyut 1 was intentionally reentered and destroyed after 175 days in orbit. The replacement unit did not reach orbit, due to a launch vehicle fault. A new mission, launched a few days before Skylab, had a failure that caused the orbit control thrusters to fire continuously, depleting the fuel supply. It reentered the atmosphere and burned.

The evolutionary architecture led to a second docking port being added. The next to last unit was abandoned and reentered due to mold problems onboard. Salyut-7 went to orbit in 1983, and was operational for more than 8 years. It was visited by ten crewed spacecraft, six with long duration crews. The Salyut series led to Mir and the ISS. The heritage of modular design for use in space was proven, and continued.

MIR

MIR ("Peace") was a Soviet modular space station, intended for long duration, Earth-orbital mission. It was launched in 1986, had visits from 28 crews, and included other nationalities. It was de-orbited in 2001. It logged the longest duration space mission to that point, more than 437 days in orbit, by two cosmonauts.

MIR was designed for a crew of 3, with hosting 6 possible for short periods. The Space Shuttle Atlantis delivered the docking module for MIR. It had seven pressurized modules, and large solar arrays for power. It was assembled in orbit, from various modules, launched separately. There were a total of seven modules, and cargo cranes to assist in maneuvering exterior equipment.

After the basic station was in orbit, 2 cosmonauts visited in 1986, and powered the station up. On the way back, they visited another orbital outpost, Salyut-7. Later, when the station was occupied, a

Kvant module was having trouble docking. An EVA revealed a trash bag in the way. Seems you can't just leave the trash outside the door, in orbit, and have it picked up.

During a docking system test in 1977, the docking vehicle collided with the Station's solar array, bounced into one of the modules, and punctured it. This lead to an emergency depressurization of the station. The crew responded quickly and contained the damage. No one was injured. In another incident, in 1997, there was an onboard fire caused by the oxygen generating equipment. This filled the facility with toxic fumes. Quick reaction by the crew resulting in containing the damage, and no injuries were reported. At the time, U. S. Astronaut Linenger was aboard. This has been the most severe fire about an orbiting space station so far. Again, lessons were learned that applied to the planned ISS. Close-out of Mir began in 1998

A resupply craft collided with the solar arrays during a test of the manual docking system. The module was holed, and was leaking the onboard atmosphere to space. Again, quick action on the part of the crew saved the day when they isolated to module by means of an airlock. Unfortunately, access to the module was lost. Later, the onboard crew was able to patch the leak, and open the airlock again.

In 1999, crews arrived to begin decommissioning the station, as the ISS project was ramping up. The station reentered and burned, landing in the South Pacific in March of 2001. The facility had been designed with a 5-year lifetime goal, but lasted 15.

MIR-II refers to a follow-on project, which eventually got incorporated into the ISS. The module used was the Zvexda ("star"). It is the basis for the current station's life support system.

Shuttle-Mir Program

The Shuttle-Mir program was a demonstration of cooperation in Space, and led to the combination of two space station programs of the United States and Russia into one facility, the International

Space Station. The U.S. wanted to take advantage of the Russian on-orbit experience, and Russia needed hard cash. This was a successful joint endeavor, continuing to this writing, regardless of the motivations.

Shuttle mission STS-63 flew to the MIR station in 1995. There was no attempt to dock; only a fly-around. There would be a total of nine Shuttle flight to Mir, delivering new modules for station expansion, crew exchange, and logistics supplies. American astronauts logged close to a thousand days onboard MIR. The station was a follow-on to the earlier Salyut Stations. It was larger and more comfortable, and took advantage of lessons-learned in orbit. With the docked Shuttle, the facility massed 250 metric tons, the world's largest facility in space.

The MIR Station was de-orbited successfully in March of 2001. It had flow three times its projected life.

ISS

In 1993, United State's Space Station Freedom Project, to create the International Space Station kicked off. On-orbit construction began in 1998, and was completed with a last Shuttle mission in 2011. It is the largest artificial satellite in Earth orbit, and can be seen from the ground with the naked eye. The ISS is a synthesis of several space station modules from the U. S., the Soviets/Russians, the Europeans,the Canadians, and the Japanese. It serves as a laboratory, observatory, and factory in Earth orbit, and is continuously crewed. Part of its mission is to collect information on items in orbit for long duration. It is currently funded through 2024. The assembly began in 1998, with the first module being the Zarya, It now has 15 pressurized modules. Five more are planned,. This is the advantage of a modular architecture. It is the 11th station sent to orbit. Early stations such as Skylab were not intended for resupply.

The International Space Station is continuously crewed, and orbits the Earth at an altitude of some 250 miles. It is quick, traveling at 17,300 miles per hour. It is also expensive, representing an investment of some $100+ billion dollars by the world community, mostly by the United States and Russia. It is thus the most expensive object ever constructed by mankind. It has been visited by astronauts and cosmonauts from some 15 nations, and by paying tourists. It can generally be found at an altitude between 300 and 435 km, and can be seen by the naked eye in the daytime, if you know here to look (there's a NASA ap for that). It has be continuously occupied for 16 years, as of this writing. It has been visited by travelers from 17 nations, some for work, some for tourism. It normally has a crew of 6, and masses 419,500 kg, the largest item in orbit.

The ISS was constructed in orbit, using large modular sections delivered to orbit by the Russian Proton heavy lift vehicle, and the Shuttle. Both had a capacity of around 50,000 pounds to Low Earth Orbit. The first section in orbit was the Zarya module.

Thirty-five Shuttle flights were used during the construction phase, to deliver components and construction crews, and serve as a temporary habitat for the crew to use during construction.

Astronauts and Cosmonauts can go out through airlocks to effect repairs to the station. This is usually hazardous. In one case, a solar array had torn. The EVA crew had to work on it when it was in sunlight, which introduced an electrocution hazard (which the suit could not have handled). The repairs were carried out carefully and successfully. In another incident, the ammonia cooling system was damaged, and one of the Astronauts sent out to evaluate and fix it got covered in ammonia, which promptly froze to his suit. In this case, he had to wait until it sublimated away (and hope his oxygen held out), because he would have contaminated the airlock with the ammonia.

The crews are rotated in and out of the station for varying periods, not to exceed a year. This service uses a Russian Soyuz capsule. There is always one capsule at the station for emergency evacuation. Each crew is referred to as an "Expedition." Current maximum is a crew of six. When a crew exchange does not involve all 3, (the capacity of the Soyuz) the spare seat may be sold to a "spaceflight participant" or space tourist. The current seat price is $40 million, includes room and board, return trip, and magnificent views. Seven have taken advantage of this opportunity to date. Cubesats coming up as cargo are accommodated by the NanoRacks system in the station, and can be attached to the outside, later to be retrieved and returned to the ground.

Now that the Shuttle fleet has been decommissioned after two disasters, the U.S. currently has no human-rated logistics vehicle, and relies on the Russian Soyuz craft for crew exchange. Crew exchange is done currently only by the Soyuz TM, but several options are in work for a new series of U. S. crewed vehicles. Several commercial companies provide non-crewed logistics services, including the Space-x (Dragon),

Logistics flights (food, oxygen science projects, clean underwear) up, and trash down now use a variety of options, the Soyuz cargo version, and commercial un-crewed capsules under contract to NASA. Some capsules burn in the atmosphere upon reentry, and some can be reused. Resupply is by Russia's *Progress* (5,200 lbs), Orbital-ATK's *Cygnus* (4400-7700) lbs, depending on launch vehicle), and SpaceX's *Dragon* (6400 lbs), the European Automated Transfer Vehicle (19,500 lbs) and the Japanese HTV Transfer vehicle (9900 lbs).

Due to orbital drag, the station needs occasional re-boosting. This can be done with any capsule docked to the Russian service module Zvezda's aft port.

For sleeping arrangements, there are no dedicated sleeping modules, and crew just typically Velcro themselves to a wall in a module of their choice. There are two Waste and Hygiene Facilities of Russian design. The liquid is piped to the Water Recovery System, and solid waste is bagged and put in a returning Progress logistics carrier.

In normal operations, the Earth's magnetic field deflects charged particles from the station. Energetic space particles may pass through the station with negligible effect. Space debris is a problem, from discarded bolts to Zombie-Sats, dead satellites in orbit. These are all tracked, and on rare occasion, the station needs to do a damage avoidance maneuver to avoid a collision. Solar flares, Coronal Mass Ejections can also endanger the crew. They then have to gather in their equivalent of a tornado shelter. A series of sentinel satellites track these events originating in the Sun, and provide ample warning time. To date, no evacuation of the Station has been necessary. The Crew normally is exposed to the same radiation in 1 day, which a person on Earth would get in a year.

And, the answer as to whether there is life in space is resoundingly "yes." By 2012, 76 types of microorganisms were detected on the station.

Decommissioning of the ISS

The current ISS will reach end-of-life in the 2020's, and is too big to be allowed to re-enter in one piece. One or more follow-on stations will be built in orbit, re-using some modules from the ISS, and new modules launched from the ground. This is possible due to the modular nature of the ISS, and the lessons-learned during its construction and use. Some of the original modules are end-of-life, and will be re-entered.

The various nations that own parts of the station are responsible for their disposal. If the parts can not be re-purposed, they will be re-entered into the atmosphere in a controlled manner. Then, the question is, what is the replacement for the ISS. More importantly, what are the follow-ons?

Russia currently has plans to remove some of its ISS modules, and re-purpose them into a new facility, the Orbital Piloted Assembly and Experiment Complex (OPSEK). This is based on an estimated life on-orbit of 30 years, based on the MIR experience.

Besides a permanently crewed station in orbit, we must begin to think of expanding outward. The near time goals are lunar and Martian. These are being addressed. These large projects will benefit from co-operation, not competition. The Deep Space Gateway Project is a joint Russian-US effort, announced in 2017.

Lessons Learned from similar Earth-based facilities

As we go further from Earth, the mission duration's will increase, and the crew will begin to feel more isolated. That has been observed in similar terrestrial scenarios such as nuclear submarines, arctic, and antarctic bases.

A nuclear submarine can stay submerged for months, with limited communications, because that can give away its position. Crews have to be trained to operate efficiency in that environment of isolation. Similarly, outposts in Antarctica and drifting ice stations at the North Pole have similar problems, but with more open communications. The research stations in Antarctica are mostly shut down over the winter, with only a small maintenance crew. This has lead to a condition called "winter-over syndrome," with a span of behavioral and medical issues. There is also a specific Polar T3 syndrome.

A specific program addressing the Mars issues is the Flashline Mars Arctic Research Station (FMARS). There is currently one such facility in the Arctic, with a second planned. The existing station is on Devon Island in the Arctic sea. It is located on a ridge, overlooking a large impact crater, about a thousand miles from the North Pole. The facility was built in 2000, and is operated by the Mars Society, a non-profit. It is used to define and refine field procedures, test habitat design, study crew performance, and selection criteria. It began operations in 2001. Generally, there is a core crew of ten, with visiting researchers and assistants. A hazard probably not found on Mars is the occasional polar bear, looking for a quick meal. One outside crew member is always armed.

Communication to and from the station to external sources is delayed 20 minutes, to simulate the one-way radio/light travel time to Mars. The crew keeps to the somewhat longer Martian Sol day

The Biosphere -2 project, located in the Arizona desert, supports 8 humans for a year in a closed ecosystem.

There has been fewer problems on the ISS and other orbiting facilities, mostly because the crew can be in constant contact, and can see home by looking out the window. The crew of a lunar backside surface facility, or heading to Mars will have to be closely monitored for indications of changes in mood and cognition.

There is also a Mars Desert Research Station in Utah in the Western United States, operated by the Mars Society. A third station is located near a volcano in Iceland. A fourth station, to be located in Australia, is in the planning stages.

Deep Space Gateway

The Deep Space Gateway (DSG) is a NASA Project for a crewed station in cis-lunar space. It is intended as a jumping-off point. The Orion crewed vehicle is scheduled to be used for this effort. The Gateway would be located in a halo orbit around the Moon. By that, we mean that the spacecraft would be visible to Earth for its entire orbital path. The DSG would form an in-orbit ecosystem for missions to the lunar surface, and to Mars. Ion thrusters are proposed for station-keeping. These use electrical power for accelerating various (usually, inert) gasses to high velocity, rather than using fuel and oxidizer. The thrust is generally low, but can be continued for long periods of time. The Gateway was going to be built for the Asteroid Redirect Mission. This mission was defunded in early 2017. It had the goal of rendezvous with an asteroid, achieving lessons-learned applicable to planetary defense.

The Power/Propulsion Module (PPM) would have large solar

arrays for power, 40 kW is baselined. The Cislunar Habitation Module will join the PPM later in orbit, and will provide living and work space. The planned Gateway and Logistics module will join next, providing experiment space and supplies. An airlock module will be added later, to enable EVA operations. The DSG can also serve as a communications relay to and from Earth, for lunar and Mars missions. Telerobotic missions on the lunar surface can be accomplished now, but the communications delay is just on the edge of making it awkward. The communications time from the DSG to the lunar surface will be negligible. The DSG could also serve as the basis for a lunar GPS system, providing a location reference for surface rovers.

Although we have data on 1-year duration mission close to Earth, the DSG will provide more information on long duration human missions in an environment away from our home planet. All of the ISS partners, US (NASA), Russia (Roscosmos), Europe (ESA), Japan (JAXA), Canada (CSA) are participating, and form the nucleus of further human exploration of the solar system.

The project presents daunting challenges in design, testing, delivery, logistics, and operations. The lunar vicinity provides a effective location for expeditions to other places in the solar system. And, it's been a while since we have stepped foot on the Moon. In the mean time, the discovery of water ice at the poles, from remote sensing, is most interesting for a lunar base.

The Gateway will serve as a "enabling infrastructure" for further exploration. It is planned to be placed in a near-rectilinear halo orbit (NRHO) around the Moon The Gateway will serve as the starting point and the mothership for lunar exploration. The DSG offers return to Earth in a matter of days. If the lunar water ice can be successfully mined and broken down into hydrogen and oxygen, we will have a fueling station that is not on the Earth's surface. With lunar surface exploration by autonomous and telerobotic rovers, mining for minerals can also be accomplished.

This will kick off with Expedition-1, an un-crewed flight of NASA's new Space Launch System (SLS) Vehicle, and Orion Spacecraft. This is scheduled, at the moment, for 2019. Various subsystems for the Orion capsule are now under test at the ISS. NASA estimates a rate of one SLS flight per year is possible, after the second flight. The basic SLS can lift 105 metric tons to Low Earth orbit, and the advanced vehicle will be capable of lifting crew and 10 metric tons to the lunar vicinity, or more than 40 tons, non-crewed.

The Gateway will differ from the ISS in several ways, but will certainly incorporated the "lessons-learned" from ISS and predecessor stations. We don't have the Shuttle anymore, so logistics will involve a new generation of heavy lift vehicles. It will probably be built in lunar orbit, so some facilities for the construction crew will be required. There is now a leaning to separate crew from cargo.
This come from the fact that everything that goes on a crewed launch vehicle has to be human-certified.
The DSG is designed to use electric propulsion for station-keeping, eliminating the need for thruster fuel, and based on the amount of available sunlight. These will be ion thrusters, that use a monopropellant.

The DSG will result in a one-year crewed mission near the moon, to validate the concept of a flight to Mars. It is not so much the distance to Mars, as the relative orbital positions of the two planets in their solar orbits. The mechanics of the transfer orbit were worked out in 1925 by German scientist Walter Hohmann. In his 1928 book, *A Daring Trip to Mars,* Max Valier shows that one of the most efficient methods of reaching Mars from Earth involves a non-intuitive Venus fly-by. People have been thinking about this for a while. The least travel time occurs about every 26 years.

Using a standard Hohmann transfer orbit would involve a 9 month travel time, 500 days at Mars, and another 9 month return journey. Due to this time-frame, there is a significant radiation risk, both in space, and on the surface, due to the thin atmosphere. There is also the issue of lack of gravity for that period of time.

The Gateway can also participate in lunar in-site research utilization, using lunar ice from the surface, brought back to the Gateway to be separated into its constituent hydrogen and oxygen, and used for rocket fuel. Hosting crewed surface missions, the Gateway would operate on a similar model to Antarctic based. U

LOP-G

The Lunar Orbital Platform - Gateway is a renaming and update to the Deep Space Gateway Program, basically a name change and some technical details updated. It will serve as staging point, in lunar orbit, for the Deep Space Transport, a re-usable crewed vehicle for Mars missions, using electrical and chemical propulsion.

The facility will have a power and propulsion element (PPE) derived from the DSG, with a mass of 8-9 metric tons, and be capable of supplying 50 KW of solar electric power for the ion thrusters. The project is in early concept phase, but some modules are being studied. These include a cis-lunar habitation module, compatible with the Orion capsule, a gateway logistics module, with a robotics arm. The Gateway airlock module will allow EVA activities, and could also be used as a short-term habitat. Lessons-learned from the ISS will be used for these modules, and certain technologies and assemblies are being tested on the station.

Follow-ons

This section discusses planned follow-ons to the LOP-G, and projects it will enable.

Deep Space Habitat

The Deep Space Habitat (DSH) was proposed in 2012, to support human exploration beyond low Earth orbit, as a stepping stone to lunar, asteroid, and Mars missions. It will utilize on-orbit experience with the ISS, and the new Orion capsule. The goal is to have a crew living and working for up to 1 year. An ideal location for the Habitat would be the L1 cis-lunar Lagrange point, a null in the Earth-Moon gravity field. The advantage of staying at this point is that both the Earth, and the Moon are "downhill" in the gravity field. It is 150 million km from Earth, and 1.5 million km from the Moon. Strangely, you can orbit a Lagrange point, even though there's nothing there. The Lagrange point on the opposite side of the Moon from the Earth is where the new James Webb Space Telescope is being placed. The five Lagrange points are solutions in the restricted three body problem of orbital mechanics. The restriction is, one of the 3 body's much be much smaller than the other two. For any two bodies, there are 5 Lagrange points. Problem is, they are not the exact null points that we would like, but are perturbed by all the other bodies in the solar system. You can trust me, or you can do the math.

The project is in Phase 2 of 3 as of this writing. Besides the Orion capsule for 4 crew members, the 60-day mission profile would utilize the Cyrogenic Propulsion System (Liquid hydrogen, liquid oxygen) assembly, a lab module, and an airlock. The MultiMission Space Exploration Vehicle (MMSEV) may also be attached. This is a servicing craft for a crew of two, and, most importantly, it includes a toilet.

A 500-day mission is also baselined, requiring the addition of a Multipurpose logistics module. The lessons learned from decades of various space station operations in orbit will be applied, as DSH will be further away from the home planet, and the cost of failures is much higher.

Russian Lunar Orbital Station

The Russian Lunar Orbital Station was a 2007 proposed project for a lunar orbital station, and an eventual surface station. It would be based on MIR and ISS lessons-learned. The project may see construction around 2030. It may also be blended into the joint US-Russian LOP-G.

NASA Lunar Outpost

The NASA Lunar Outpost is an element of the Bush administration (2001-2009) *Vision for Space Exploration*. It was directed by Congress that the facility would be named the *Neil A. Armstrong Lunar Outpost*. The outpost location would be at one of the lunar poles. Since then, remote observations have revealed the presence of water ice in craters at the South Pole The South Poles remains in shadow, and sunlight does not reach the bottom of the craters. Besides the value of in situ water supplies, and the ability to produce hydrogen and oxygen from the water via solar-powered hydrolysis, the ice may contain records of the material of the early solar system. The water is a critical sustainable resource for a crewed lunar base. The Indian Chandrayaan lunar orbiter was responsible for this discovery.

One ideal structure for lunar bases (that the author has worked on) is lava tubes. These are found on the Earth as well, and the cooled lava provides a hard, sealed surface. It just needs to be capped with airlock doors, and no further exterior construction is required.

The outpost will consist of various modules for habitation and laboratory space, an extension solar array assembly, and a garage for a rover. There will also be a communications facility for the link to Earth. The facility was designed for a crew of four with 7-day visits during deployment, and up to 180 day operational missions.

NASA is looking to private enterprise to provide logistics support for the lunar surface, and some companies are making lunar plans themselves. Bigelow Aerospace has plans to build bases on the moon, based on their inflatable modules, originally developed from NASA technology. There is one of these attached to the ISS right now.

The Google Lunar X-Prize contest was not completed successfully by any team, and has been discontinued, Privately funded teams have to land a robotic spacecraft on the lunar surface, travel at least 500 meters, and transmit back high-definition video and pictures. Five Teams were still in the running out of 18, at the deadline. All of the teams are developing relevant lunar technology. The winner could have received as much as $30 million.

At the moment, NASA is not planning to send any crew to the lunar surface, but is focusing on the LOP-G, and an eventual Mars mission.

Transportation and Logistics

The radiation environment of Earth orbit is well understood. The crew on the ISS can operate for up to a year at the ISS, before accumulating a "life-time" dose. In additional Sentinel satellites give us a few days warning of solar storms, or Coronal Mass Ejections, that would allow the crew to enter a specially designed "storm shelter" for a few days. The situation at the moon is much different. The moon has a weak magnetic field compared to Earth. The good news is, there is almost no trapped particles, like Earth's Van Allen Belts. The bad news is, the lack of an appreciable magnet field mean no protection from energetic charged particles. In cis-lunar space, there is a greater incidence of galactic cosmic rays. It is a similar situation at Mars. For Habitats at the Earth-Moon libration points, it is the same. Facilities on the lunar surface will probably be put under a layer of regolith. Another potential accommodation is within lunar lava tubes.

Away from our home planet with its convenient magnetic field and van Allen Belts, we have radiation issue with cosmic rays and the Solar proton wind, as well as transient events such as the CME. This will be addressed by locally obtained mass, from the lunar surface or asteroid, employed as bulk shielding. The shielding will also protect against space debris. Although a window may be pierced, the hole will generally be small enough to be ignored for a while, due to the large enclosed volume of air.

Colonies are best located near resources and energy sources. The energy resource is the Sun. The resources, in the short term are the moon and the asteroids. This approach was followed in 19th century iron manufacturing, where the iron furnaces were located near supplies of the raw materials, iron ore, limestone, and coal. Villages grew up around the iron works. It is cheaper to ship finished product than the raw materials.

Water ice is a valuable commodity. It has been observed in craters at the lunar south pole. It can be electrolized with abundant solar power into its constituent hydrogen and oxygen, and used as rocket fuel.

There also seems to be extensive amounts of Helium-3 in the lunar regolith, from the solar wind. Helium-3 is a potential energy source in nuclear fusion reactors. It has the nice property of releasing large amounts of energy, but little radiation.

Besides the asteroids, which mainly are found beyond Mars, there are also inactive or captured comets that contain water ice, and hydrocarbons. Jupiter's Trojan asteroids may have water ice in vast quantities, which would make them ideal filling stations for interplanetary missions. Just need a bathroom, coffee machine, and some snacks.

Energy from the Sun is abundant, and amounts to about 1367 watts per square meter, at Earth's orbit. More as you go closer, less as you go further away, by an inverse square law. With current solar cell technology, solar panels can be used up to Jupiter, but not beyond.

Mars Base Camp

Human missions to Mars have gotten consideration for more than 150 years. Von Braun made a very detailed study of a Mars mission in 1952. Willy Ley had published a variation in 1949. By 1956, their modified mission design would required over 400 launches, assembly in space, but would provide a winged lander for Mars. The thought is to land there, and, now that we know the surface features, begin to terraform the planet for our needs. Mars missions are in active planning by the United States, Russia, Europe, China, and several commercial entities. The best transfer orbit was defined in 1925 by Walter Hohmann. Since Mars and the Earth have different orbital periods ("years") around the Sun, there are optimal times to make the journey. After working through all the math, the time between optimal Earth-Mars trips is 26 months. This provides the optimal energy expenditure. Using a Hohmann transfer, there would be a 9-month travel time, Earth to Mars, a 500 day stay at Mars to allow alignment of the orbits again, and another 9 month journey back. Can't we do better, yes, but at the coast of fuel. There is a maneuver that would provide a Venus and a Mars flyby in one mission, with no landings. All of these long duration missions offer various hazards, and will be operating in "unknown territory" in terms of a small group of humans in a confined space for a long time.

A NASA study by three major aerospace companies in 1962, showed a Mars mission requiring 8 Saturn-V "moon rockets," with assembly in Earth orbit. Von Braun's Mars Mission was passed over in favor of the Space Shuttle Project. Mars has been studied by fly-by, orbiting, and lander missions since 1981. NASA's Mars Design Reference Mission of the 1990's assumed that fuel could be synthesized from Martial atmospheric or surface components. The design studies continue to this day.

The Mars Base Camp is a crewed Mars orbiter, proposed by Lockheed Martin, possibly ready for the 2028 favorable launch opportunity. The hardware would use the Orion MPCV. Although the humans would remain in Mars orbit, they would perform tele-operation experiments on the surface. The mission would be launched from lunar orbit. The first mission is named Mars Base Camp-1. A crew of 6 would spend a year in Martian orbit. The duration is dictated by the relative orbital positions of Mars and Earth. Interplanetary missions generally provide very limited options for abort/early return scenarios.

Lockheed defined a road map of the technologies required to achieve the Mission. The parts include: the MPCV, which is the orbital command and control center, implementing navigation, communications, and life support and habitat systems. The Solar Array subsystem provides power for the system, including the electric propulsion engines. This is being developed by NASA-Glenn. The radiators control dumping of excess heat. The propulsion stage uses cyrogenic fuels. Side visits to the moons Phobos and Deimos are planned. The main part of the assembly will consist of an Orion capsule with an associated service module, and an excursion module. There will be a laboratory and workshops. The excursion modules will provide access to the Martian (and Mars lunar surfaces.)

The Mars hardware and operations would be checked out at the LOP-G. The eventual goal is to establish a self-sustaining colony, and perhaps terraform the Red Planet to be friendlier to Earth life, plant (potatoes), and animals.

The MPCV would provide housing, life support, transportation, and command & control. Large solar arrays would be used, as the vehicle is designed for electric propulsion. In addition, an onboard 3-D printer is seen as an alternative to manifesting spares.

A follow-on concept is winged Mars surface lander, using liquid hydrogen and liquid oxygen engines. The fuel and oxidizer would be derived from water, electrolized using solar energy. Initially, the water would be shipped from Earth or the Moon, with a goal of finding water on the Martian surface, or its moons. NASA sees an opportunity for commercial firms to supply the water. A Water Delivery Vehicle (WDV) with a capacity of over 50 metric tons would be required. This will dock with a 375 kilowatt electrolysis plant in orbit. The Earth-rated 375 kw plant degrades to 160kW (42.6%) at Mars' greater distance from the Sun.

Modules will be delivered and pre-positioned in Mars orbit, including a lander (The Mars Ascent/Descent Vehicle) and a cyrogenic fuel depot. The MADV would touche down and launches vertically. It will house the crew for a 10-day duration on the surface. The MADV's engines will use liquid hydrogen and oxygen for an efficient Isp of 405 seconds. Six RL-10 class engines are postulated.

The lander will have crew accommodations for four, on three decks. There is to be a flight deck, crew quarters on the mid deck, and an aft deck with galley, lab facilities, and the airlock.

Solar electric propulsion using xenon gas will be feasible with large solar arrays. The LOP-G will use this approach for station-keeping.

One timeline shows a cis-lunar outpost starting in 2021, with lunar surface science by 2024. There could be a pre-deployment of assets to Mars by 2026. The year 2028 is targeted for the Mars mission. Besides the government mission, certain private spacecraft companies are defining their own Mars missions.

Commercial ventures such as Elon Musk's SpaceX are also interested in a Mars Colony. Musk, the founder of PayPal, put his own money into the company. It has developed its own launch vehicle, the Falcon, and has a 12-trip resupply contract for the ISS. SpaceX's approach to the un-crewed resupply missions is unique. The capsule is recovered and reused, saving a lot of money.

Broader Space Colonization

As of this writing, the space activities have gone from an Astronaut on a ballistic trajectory, to a fully crewed multi-year space station, to plans for the LOP-G, lunar bases, and Mars missions.

The next steps have been known for decades – permanent habitation on another planet. That, and human exploration of the outer planets will take our attention for decades into the future. After that, the solar system won't be able to hold us. This will require technologies that have yet to be developed. And, throughout history, humans were explorers. They always wondered "what was over there?"

In the shorter term, space Colonization will take our best efforts for a long time. It is obvious that the colonists will need to utilize in-situ materials and energy sources. Space is hostile to life. That's why our planet is important to us.

Why are the expense and effort? For one thing, it might help to ensure the survival of our civilization. If there was a global disaster, having friends in other places might help us recover. That could include direct assistance, and additional resources.

Space is full of vast resources for us to utilize. We just need the technology to do it. We are not going to be able to lift all the materials we need from the Earth's surface, so it will be important to develop the technologies for lunar and asteroid mining and processing. Many Earth resources are non-renewable, but could be replaced with off-planet resources. Let's plan from the beginning not to trash other planets. Resources that can't be used, or are radioactive or toxic, can be sent on a one-way trip to the Sun.

Orbital Development Corporation is proposing a project to exploit the near-Earth asteroid Eros. It's some 20 miles long, with an estimated mass of nearly 80 trillion tons (10^{12}). Thanks to the NEAR Shoemaker spacecraft, much of its composition is known, and includes iron, aluminum, silicon, and magnesium. It is likely that rare-earth metals are also present. It is proposed to build a habitat at one point on the spin axis, with a shipyard at the other. Mr. Gregory W. Nemitz claims ownership of the asteroid, which has been contested in Federal Court. He sent NASA a parking bill for the NEAR spacecraft on the surface. We'll see how this all works out, but it is precedent-setting, and needs to be decided before large-scale commercial exploitation of space resources is begun.

(http://www.orbdev.com/erosproj.html)

Wrap-up

The Lunar Orbital Platform - Gateway, the lunar outpost, the Mars Mission. These are no longer concepts, but active projects with a lot of smart people, and a lot of money. It's not a question of whether its going to happen, but how soon. With Commercial firms involved and interested in mining the moon and asteroids, Earth will have to develop more complex "Space Law" to address who owns what and who benefits. In the past, the new frontiers, America, the Yukon, the "West" were mostly wild and ungoverned, at least at first. Hopefully, we will think this thing through, so no corporation or Nation-state will be able to enrich themselves, at the cost of others.

As of this writing, several NASA contracts have been let to Industry to implement the LOP-G. Maxar Technologies, of Coloradois to "develop and demonstrate power, propulsion, and communications capabilities." Maxar used to be Space Systems Loral.

The power and propulsion unit is a 50 kw electric propulsion The contract is worth $375. million dollars. The contract period includes a flight demonstration of up to a year, with Maxar retaining ownership.

This is happening. Mankind is going back to the Moon.

Bibliography

Aldrin, Buzz *No Dream Is Too High, Life Lessons From a Man Who Walked on the Moon*, National Geographic, 2016, ISBN-9781426216497.

Bell, S. S.(ed), McCoy, Earl D. (ed), Mushinsky, H.R. (Ed) *Habitat Structure: The physical arrangement of objects in space*, 1990, ISBN-0412322706.

Benaroya, Haym *Building Habitats on the Moon: Engineering Approaches to Lunar Settlements,* (Springer Praxis Books), 2018, ISBN-3319682423.

Buckley, James *Home Address: ISS: International Space Station,* Smithsonian, Sep 1, 2015.

Burns, Jack O. Terry Fong NASA Ames Research Center David A. Kring William D. Pratt and Timothy Cichan, "SCIENCE AND EXPLORATION AT THE MOON AND MARS ENABLED BY SURFACE" TELEROBOTICS" INTERNATIONAL ACADEMY OF ASTRONAUTICS 10th IAA SYMPOSIUM ON THE FUTURE OF SPACE EXPLORATION: TOWARDS THE MOON VILLAGE AND BEYOND," Torino, Italy, June 27-29, 2017.

Buss, Jared S. *Willy Ley: Prophet of the Space Age*, 2017, University Press of Florida, ISBN-0813054435.

Chladek, Jay *Outposts on the Frontier: A Fifty-Year History of Space Stations (Outward Odyssey: A People's History of Spaceflight*, 2017 U. Nebraska Press, ISBN-0803222920.

Cichana, Timothy; O'Dellb, Sean; Richeyc, Danielle; Baileyd, Stephen A.; Burche, Adam "MARS BASE CAMP UPDATES AND NEW CONCEPTS," 2017, IAC-17, 68th International Astronautical Congress (IAC).

Cohen, Marc *Space Architecture* (Advances in Engineering Series), 2016, ISBN-1466505451.

de Vera, Jean-Pierre, Seckbach, Joseph *Habitability of Other Planets and Satellites* (Cellular Origin, Life in Extreme Habitats and Astrobiology), 2013, ISBN-3540223150.

DoD, *21st Century Essential Guide to Navy Submarines: Past, Present, and Future of the Sub Fleet, History, Technology, Ship Information, Pioneers, Cold War, Nuclear Attack, Ballistic, Guided Missile,* 2011, ASIN-B004OEIMX0.

Eckhart, Peter *The Lunar Base Handbook,* 1999, 1st ed, McGraw-Hill Primis Custom Publishing, ASIN-B01A1MSBRK.

Ferraris, Silvia Deborah, *Living in space: Industrial Design contribution to the habitat quality of spacecrafts*, 2010, LAP LAMBERT Academic Publishing, ISBN-103838377222.

Griffin, Brand "Skylab II: Making a Deep Space Habitat from a Space Launch System Propellant Tank, "March 27, 2013 Future In-Space Operations Colloquium, Future In-Space Operations Working Group. Avail: http://spirit.as.utexas.edu/~fiso/telecon13-15/Griffin_3-27-13/.

Fowler, Wallace *Moon Port: Transportation Node in Lunar Orbit: NASA's Effort to Support a Manned Lunar Colony*, 2015, ASIN – B014LRVBOQ.

Gatens, Robyn; Crusan, Jason "Cislunar Habitation & Environmental Control & Life Support System" available as a .pdf, www.nasa.gov.

Hale, Edward Everett *The Brick Moon and Other Stories*, 1869, reprint 2011, ASIN-B004TPHETM.

Häuplik-Meusburger, Sandra Olga Bannova, Olga *Space*

Architecture Education for Engineers and Architects: Designing and Planning Beyond Earth (Space and Society), 2016, Springer, ISBN-9783319192796.

Heimreich, Robert L. "The undersea habitat as a space station analog evaluation of research and training potential" (SuDoc NAS 1.26:180342), 1985.

Heppenheimer, T. A. *Colonies in Space, A Comprehensive and Factual Account of the Prospects for Human Colonization of Space*, 1977, ISBN-0-446-81-581-0.

Ley, Willy *Space Stations: Adventures in Space*, 1958, Guild Press, ASIN-B000LB5OMC.

Ley, Willy, Rockets, *The Future of Travel Beyond the Stratosphere*, 1945, Viking Press, ASIN- 0007E7IC2.

Lockard, Elizabeth Song *Human Migration to Space: Alternative Technological Approaches for Long-Term Adaptation to Extraterrestrial Environments*, 2014, Springer, ISBN-3319059297.

Mendell, Wendell W. *Lunar bases and Space Activities of the 21st century*, 1985, Lunar and Planetary Institute, ISBN 0-942862-02-3.

Merrow, Mark *A Lunar Space Station: NASA's Study to Design a Lunar Space Station in Support of a Manned Moon Base,* 2015, alc Books, ASIN-B014LQ177S.

NASA, Office of Inspector General, "NASA's Plans for Human Exploration Beyond Low Earth Orbit," 2017, Report IG-17-017, avail: https://oig.nasa.gov/audits/reports/FY17/IG-17-017.pdf

NASA, *Inside the International Space Station (ISS): NASA International Space Station Familiarization Astronaut Training Manual - Comprehensive Review of ISS Systems*, 2011, ASIN-

B006O403MG.

NASA, *NASA Report on Mars Exploration: Frontier In-Situ Resource Utilization for Enabling Sustained Human Presence on Mars - ISRU, Surface Habitats, Entry Descent and Landing, Fuels, Food, Robotics,* 2016, ASIN-B01JDORX58.

NASA, *NASA Space Technology Report: Lunar and Planetary Bases, Habitats, and Colonies, Special Bibliography Including Mars Settlements, Materials, Life Support, Logistics, Robotic Systems,* ASIN-B00CLX44E2.

NASA, *NASA Space Technology Report: Deep Space Habitat Concept of Operations for Transit Mission Phases - Mars, Phobos / Deimos, Near Earth Asteroid, Habitats, Crew Systems,* 2013, ASIN-B00EG4N3E6.

"L1 libration point manned space habitat," NASA-USRA advanced space mission design project" (SuDoc NAS 1.26:184732), 1987.

Nixon, David *International Space Station: Architecture beyond Earth*, 2017, Circa Press, ISBN 0993072135.

Noordung, Hermann; Potocnik, Herman *The Problem of Space Travel: The Rocket Motor,* 1928, reprint 2015, ASIN-B015EYQR9O.

Oberth, Hermann *MENSCHEN IM WELTRAUM: NEUE PROJEKTE FÜR RAKETEN- UND RAUMFAHRT, ("People in space – New projects for rockets and space travel"),* 1957, ASIN-B0000BM2NL.

O'Neill, Gerald K.; Dyson, Freeman *The High Frontier: Human Colonies in Space*: Apogee Books Space Series 12,3rd Edition, 2000, ISBN-189652267X.

O'Neill, Gerald K. *2081*, Space Studies Institute, 2017, ASIN-

B0759STNPL.

Popular Science (ed) *The Future of Space Travel: Your New Ride to Space, 2017,* ISBN-1683308182.

Portree, David S. F. *Humans to Mars: Fifty Years of Mission Planning, 1950–2000,* NASA Monographs in Aerospace History Series, Number 21, February 2001, NASA SP-2001-4521. Avail: ASIN-B014RGH7GM.

Powell-Willhite, Irene E. *The Voice of Dr. Wernher von Braun*: *An Anthology,* 2007, Collectors Guide Publishing, ISBN-1894959647.

Rapp. Donald *Human Missions to Mars: Enabling Technologies for Exploring the Red Planet,* 2015, Springer Praxis, ISBN-3319222481.

Schrunk, David; Sharpe, Burton *The Moon: Resources, Future Development and Settlement* (Springer Praxis Books), 2007, ISBN-0387360557.

Schwartz, James S. J.; Milligan, Tony *The Ethics of Space Exploration,* Springer, 2016, ISBN-3319398253

Seedhouse, Erik *Bigelow Aerospace: Colonizing Space One Module at a Time,* 2015, Springer, ISBN-3319051962.

Seedhouse, Erik *Lunar Outpost: The Challenges of Establishing a Human Settlement on the Moon,* 2008, Springer, ISBN-0387097465.

Smitherman, David *Habitat Concepts for Deep Space Exploration,* 2014, NASA, ASIN-B01ED7JF10.
Stapleton, Olaf, *Star Maker,* 1937, ISBN-1-85798-807-8.

Tsiolkovsky, Konstantin E. *Selected Works of Konstantin E.*

Tsiolkovsky, 2004, University Press of the Pacific, ISBN-141021825

United States Congress and United States House of Representatives, *Next step to Mars: deep space habitat* : hearing before the Subcommittee on Space, 2017.

Valier, Max; Miller, Ron (ed), *A Daring Trip to Mars*, 1928, reprint, 2013, ASIN-B00CSWANK0.

Von Braun, Werhner *Project MARS, a Technical Tale*, 1971, ISBN-0-9738203-3-0.

Von Braun, Werhner *The Mars Project*, 1962, U. Illinois Press, ISBN-0252062272.

Whitley, Ryan; Martinez, Roland; Condon, Gerald; Williams, Jacob; Lee, David; Davis, Diane; Barton, Gregg; Bhatt, Sagar; Jang, Jiann-Woei; Clark, Fred "Cislunar Near Rectilinear Halo Orbit for Human Space Exploration," 2016, NASA NTRS, Report 20160003078.

Zubrin, Robert *Mars on Earth: The Adventures of Space Pioneers in the High Arctic*, 2003, ISBN-158542255X .

Zubrin, Robert *Entering Space: Creating a Spacefaring Civilization*, 2000, ISBN-10-1585420360.

Zubrin, Robert *Mars Direct: Space Exploration, the Red Planet, and the Human Future: A Special from Tarcher/ Penguin*, 2013, ASIN-B00AMOO98I.

Resources

- NASAspaceflight.com
- https://www.nasa.gov/feature/deep-space-gateway-to-open-opportunities-for-distant-destinations

- https://www.reddit.com/r/spacex/comments/623pmx/dragon_2_as_commercial_replacement_for_orion_crew/
- https://www.nasa.gov/pdf/315828main_LSS_Overview_for_Industry_Culbert.pdf
- https://www.nasa.gov/pdf/240373main_06-06-08-LSS%20BAA%20Compilation%20of%20briefings%20teammb.pdf
- http://www.lockheedmartin.com/us/ssc/mars-orion.html
- Taking out the trash: https://arc.aiaa.org/doi/abs/10.2514/6.2017-5126
- Space Studies Institute, www.ssi.org
- David Hardy, PROJECT HYPERION: THE HOLLOW ASTEROID STARSHIP, avail: http://www.icarusinterstellar.org/project-hyperion-the-hollow-asteroid-starship-dissemination-of-an-idea/
- Mars Base Camp, http://lockheedmartin.com/us/ssc/mars-orion.html
- "Mars Base Camp Updates and New Concepts" available for download at the address above.
- NASA's Exploration Systems Architecture Study -- Final Report, avail: https://www.nasa.gov/exploration/news/ESAS_report.html
- Human Exploration of Mars, Reference Mission avail:https://web.archive.org/web/20070626154441/http://exploration.jsc.nasa.gov/marsref/contents.html
- wikipedia, various.

Glossary of terms and definitions

Apogee – furthest point in the orbit from the Earth.
Aphelion – furthest point to the Sun.
Apolune – furthest point to the Moon.
ARCM – asteroid redirect crewed mission.
ARM – asteroid redirect mission
ARRM – asteroid retrieval robotic mission
ASIN – Amazon Standard Inventory Number
Astrionics – electronics for space flight.
BEAM – Bigelow Expandable Activity Module – commercial inflatable space module.
BEO – beyond Earth orbit.
CATALYST - Lunar Cargo Transportation and Landing by Soft Touchdown
CBM – common berthing mechanism
CHM – cis-lunar habitation module
Cislunar – beyond Earth's atmosphere to just beyond the moon's orbit.
CM – crew module
CME – Coronal Mass Ejection, blast of energetic particles from the Sun.
CMP – co-manifested payload.
CNSA – China National Space Administration.
Conops – concept of operations.
CPS – Cyrogenic Propulsion Stage.
CRTBP – Circular Restricted three-body Problem.
CSA – Canadian Space Agency, Agence Spatiale Canadienne
CSF – Cislunar Support Flight.
C&W – caution and warning.
Cygnus – Orbital-ATK automated cargo vehicle for ISS.
Cyrogenic – relating to very low temperatures.
DAM – damage avoidance maneuver.
DCM – docking cargo module.
Delta-V – change in velocity.
DoD – (U.S.) Department of Defense

DRG – Distant Retrograde Orbit.
DRM – design reference mission.
DRO – distant retrograde orbit.
DSG – Deep Space Gateway
DSH – deep space habitat.
DSN – (NASA) Deep Space Network.
DST – Deep Space Transport
DTM – dynamic test model, for structural tests.
ECLSS – Environmental Control & Life Support system.
EDL – Entry, Descent, Landing.
EM-x Exploration Mission number-x.
Ephemeris – position information data set for orbiting bodies, 6 parameters plus time.
Epoch – a reference point in time for orbital elements.
EPS – electrical power system
ESA – European Space Agency
EUS – Exploration Upper Stage.
EVA – extra-vehicular activity.
FMARS – Flashline Mars Arctic Research Station
GAM – Gateway Airlock Module
GLM – Gateway Logistics Module
GNC – Guidance, Navigation, and Control.
Gravity well – a conceptual model of the gravity field near a mass.
GSFC – NASA Goddard Space Flight Center, Greenbelt, MD.
Halo Orbit – three dimension orbit near the L1, L2, or L3 Lagrange points.
HEEO – highly eccentric Earth orbit.
HEOMD – Human Exploration and Operations Mission Directorate.
HITL – Human in the loop.
HOPE – Human Outer Planet Exploration (NASA)
HSIR – human systems integration requirements
IDSS – International Docking System Standard.
IGA - (ISS) InterGovernmental Agreement
ISP – specific impulse. Measure of efficiency of rocket engine. Units of seconds.
ISRO – Indian Space Research Organization

ISRU – in situ resource utilization
ISS – International Space Station
JAXA – Japan Aerospace Exploration Agency.
KW – kilowatt.
ISRU – in site resource utilization.
ISS – International Space Station
JAXA – Japanese space agency
JPL – Jet Propulsion Laboratory, Pasadena, CA.
JSC – Johnson Space Center, Houston, Texas.
KSC – NASA Kennedy Space Center, launch site, Florida.
L2 – second of 5 Lagrange points, a null in the gravity field in the restricted 3-body problem.
LAS – launch abort system
Lbf – pounds, force.
LCT – Lunar Cargo Transportation.
LEO – Low Earth Orbit
LH2 – liquid hydrogen.
Libration point – null in the gravity field of the three body problem.
LOS – Russian Lunar Orbital Station;m loss-of-signal.
LOX – liquid oxygen, boils at -297 F.
LSAM – lunar surface access module
LSSPO – Lunar Surface Systems Project Office (NASA-JSC).
LST – landing by soft touchdown.
MADV – Mars Ascent/Descent Vehicle.
MBC – Mars Base Camp.
MET – mission elapsed time.
MMSEV – MultiMission Space Exploration Vehicle.
MOU – memorandum of understanding.
MPCV - Multi-Purpose Crew Vehicle.
MPLM – Multi-purpose Logistics Module.
m/s – meters per second.
Mt – metric ton, 1000 kg.
NAC – NASA Advisory Council.
Nadir – the point directly below.
NASA – (U.S.) National Aeronautics and Space Administration
NEO – near Earth object.

NextSTEP-2 – (NASA) Next Space Technologies of Exploration Partnerships.
NHV – net habitable volume.
NRHO – Near rectilinear halo orbit (around the L1 or L2 Earth-Moon libration point).
NTIS – National Technical Information Service (www.ntis.gov).
NTRS – NASA Technical Reports Server, (ntrs.nasa.gov)
ORU – Orbital Replacement Unit.
OPSEK – (Russian) Orbital Piloted Assembly and Experiment Complex.
Perigee –closest point in the orbit from the Earth.
Perhelion – closest point to the Sun.
Perilune – closest point to the Moon
PMA – Pressurized mating adapter.
PMCU – Power Management Control Unit.
PPB – power and propulsion bus
PPE – power and propulsion element
PTCS – Passive thermal control system
PVCU – Photo Voltaic Control Unit.
RCS – reaction control system.
RGA – rate gyro assembly
R&D – research & development.
Regolith – layer of loose material, covering rock; dirt.
ROSCOSMOS – Russian Space Agency.
RPOD – Rendezvous, Proximity Operations, Docking.
SEP – solar electric propulsion
SHFE – space human factors engineering.
SI – System International – the metric system.
Sidereal period – time for an object to make a full orbit.
Sol, local solar day – on Mars, 24h, 37 min.
SLS – (NASA) Space Launch System.
SPACE Act - Spurring Private Aerospace Competitiveness and Entrepreneurship
Synodic period - time for an object in orbit to occupy the same point, in relation to 2 other objects.
TCS – thermal control system.
TLI – Trans-lunar injection.

TM – Technical Manual.
TPS – thermal protection system.
Trillion - 10^{12}
TRL – technology readiness level.
UDM – universal docking module.
Ullage – residual fuel or oxidizer in a tank after engine burn is complete.
USAF – United States Air Force.
V&V – verification and validation.
WDV – water delivery vehicle.
XBASE - Expandable Bigelow Advanced Station Enhancement.
Zenith – the point directly above.
Zombie-sat – a non functional satellite in orbit, contributing to the orbital debris problem.

If you enjoyed this book, you might also be interested in some of these.

Stakem, Patrick H. *16-bit Microprocessors, History and Architecture*, 2013 PRRB Publishing, ISBN-1520210922.

Stakem, Patrick H. *4- and 8-bit Microprocessors, Architecture and History*, 2013, PRRB Publishing, ISBN-152021572X,

Stakem, Patrick H. *Apollo's Computers,* 2014, PRRB Publishing, ISBN-1520215800.

Stakem, Patrick H. *The Architecture and Applications of the ARM Microprocessors,* 2013, PRRB Publishing, ISBN-1520215843.

Stakem, Patrick H. *Earth Rovers: for Exploration and Environmental Monitoring,* 2014, PRRB Publishing, ISBN-152021586X.

Stakem, Patrick H. *Embedded Computer Systems, Volume 1, Introduction and Architecture*, 2013, PRRB Publishing, ISBN-1520215959.

Stakem, Patrick H. *The History of Spacecraft Computers from the V-2 to the Space Station*, 2013, PRRB Publishing, ISBN-1520216181.

Stakem, Patrick H. *Floating Point Computation*, 2013, PRRB Publishing, ISBN-152021619X.

Stakem, Patrick H. *Architecture of Massively Parallel Microprocessor Systems*, 2011, PRRB Publishing, ISBN-1520250061.

Stakem, Patrick H. *Multicore Computer Architecture,* 2014, PRRB

Publishing, ISBN-1520241372.

Stakem, Patrick H. *Personal Robots*, 2014, PRRB Publishing, ISBN-1520216254.

Stakem, Patrick H. *RISC Microprocessors, History and Overview,* 2013, PRRB Publishing, ISBN-1520216289.

Stakem, Patrick H. *Robots and Telerobots in Space Applications*, 2011, PRRB Publishing, ISBN-1520210361.

Stakem, Patrick H. *The Saturn Rocket and the Pegasus Missions, 1965,* 2013, PRRB Publishing, ISBN-1520209916.

Stakem, Patrick H. *Visiting the NASA Centers, and Locations of Historic Rockets & Spacecraft,* 2017, PRRB Publishing, ISBN-1549651205.

Stakem, Patrick H. *Microprocessors in Space*, 2011, PRRB Publishing, ISBN-1520216343.

Stakem, Patrick H. Computer *Virtualization and the Cloud*, 2013, PRRB Publishing, ISBN-152021636X.

Stakem, Patrick H. *What's the Worst That Could Happen? Bad Assumptions, Ignorance, Failures and Screw-ups in Engineering Projects, 2014,* PRRB Publishing, ISBN-1520207166.

Stakem, Patrick H. *Computer Architecture & Programming of the Intel x86 Family, 2013,* PRRB Publishing, ISBN-1520263724.

Stakem, Patrick H. *The Hardware and Software Architecture of the Transputer*, 2011,PRRB Publishing, ISBN-152020681X.

Stakem, Patrick H. *Mainframes, Computing on Big Iron*, 2015, PRRB Publishing, ISBN- 1520216459.

Stakem, Patrick H. *Spacecraft Control Centers*, 2015, PRRB Publishing, ISBN-1520200617.

Stakem, Patrick H. *Embedded in Space,* 2015, PRRB Publishing, ISBN-1520215916.

Stakem, Patrick H. *A Practitioner's Guide to RISC Microprocessor Architecture*, Wiley-Interscience, 1996, ISBN-0471130184.

Stakem, Patrick H. *Cubesat Engineeering*, PRRB Publishing, 2017, ISBN-1520754019.

Stakem, Patrick H. *Cubesat Operations*, PRRB Publishing, 2017, ISBN-152076717X.

Stakem, Patrick H. *Interplanetary Cubesats*, PRRB Publishing, 2017, ISBN-1520766173 .

Stakem, Patrick H. Cubesat Constellations, Clusters, and Swarms, Stakem, PRRB Publishing, 2017, ISBN-1520767544.

Stakem, Patrick H. *Graphics Processing Units, an overview*, 2017, PRRB Publishing, ISBN-1520879695.

Stakem, Patrick H. *Intel Embedded and the Arduino-101, 2017,* PRRB Publishing, ISBN-1520879296.

Stakem, Patrick H. *Orbital Debris, the problem and the mitigation*, 2018, PRRB Publishing, ISBN-*1980466483*.

Stakem, Patrick H. *Manufacturing in Space*, 2018, PRRB Publishing, ISBN-1977076041.

Stakem, Patrick H. *NASA's Ships and Planes*, 2018, PRRB Publishing, ISBN-1977076823.

Stakem, Patrick H. *Space Tourism*, 2018, PRRB Publishing, ISBN-

1977073506.

Stakem, Patrick H. *STEM – Data Storage and Communications*, 2018, PRRB Publishing, ISBN-1977073115.

Stakem, Patrick H. *In-Space Robotic Repair and Servicing*, 2018, PRRB Publishing, ISBN-1980478236.

Stakem, Patrick H. *Introducing Weather in the pre-K to 12 Curricula, A Resource Guide for Educators*, 2017, PRRB Publishing, ISBN-1980638241.

Stakem, Patrick H. *Introducing Astronomy in the pre-K to 12 Curricula, A Resource Guide for Educators*, 2017, PRRB Publishing, ISBN-198104065X.
Also available in a Brazilian Portuguese edition, ISBN-1983106127.

Stakem, Patrick H. *Deep Space Gateways, the Moon and Beyond*, 2017, PRRB Publishing, ISBN-1973465701.

Stakem, Patrick H. *Exploration of the Gas Giants, Space Missions to Jupiter, Saturn, Uranus, and Neptune*, PRRB Publishing, 2018, ISBN-9781717814500.

Stakem, Patrick H. *Crewed Spacecraft*, 2017, PRRB Publishing, ISBN-1549992406.

Stakem, Patrick H. *Rocketplanes to Space*, 2017, PRRB Publishing, ISBN-1549992589.

Stakem, Patrick H. *Crewed Space Stations*, 2017, PRRB Publishing, ISBN-1549992228.

Stakem, Patrick H. *Enviro-bots for STEM: Using Robotics in the pre-K to 12 Curricula, A Resource Guide for Educators*, 2017, PRRB Publishing, ISBN-1549656619.

Stakem, Patrick H. *STEM-Sat, Using Cubesats in the pre-K to 12 Curricula, A Resource Guide for Educators*, 2017, ISBN-1549656376.

Stakem, Patrick H. *Lunar Orbital Platform-Gateway*, 2018, PRRB Publishing, ISBN-1980498628.

Stakem, Patrick H. *Embedded GPU's*, 2018, PRRB Publishing, ISBN- 1980476497.

Stakem, Patrick H. *Mobile Cloud Robotics*, 2018, PRRB Publishing, ISBN- 1980488088.

Stakem, Patrick H. *Extreme Environment Embedded Systems,* 2017, PRRB Publishing, ISBN-1520215967.

Stakem, Patrick H. *What's the Worst, Volume-2*, 2018, ISBN-1981005579.

Stakem, Patrick H., *Spaceports*, 2018, ISBN-1981022287.

Stakem, Patrick H., *Space Launch Vehicles*, 2018, ISBN-1983071773.

Stakem, Patrick H. *Mars*, 2018, ISBN-1983116902.

Stakem, Patrick H. *X-86, 40th Anniversary ed*, 2018, ISBN-1983189405.

Stakem, Patrick H. *Lunar Orbital Platform-Gateway*, 2018, PRRB Publishing, ISBN-1980498628.

Stakem, Patrick H. *Space Weather*, 2018, ISBN-1723904023.

Stakem, Patrick H. *STEM-Engineering Process*, 2017, ISBN-1983196517.

Stakem, Patrick H. *Space Telescopes,* 2018, PRRB Publishing, ISBN-1728728568.

Stakem, Patrick H. *Exoplanets*, 2018, PRRB Publishing, ISBN-9781731385055.

Stakem, Patrick H. *Planetary Defense*, 2018, PRRB Publishing, ISBN-9781731001207.

Patrick H. Stakem *Exploration of the Asteroid Belt*, 2018, PRRB Publishing, ISBN-1731049846.

Patrick H. Stakem *Terraforming*, 2018, PRRB Publishing, ISBN-1790308100.

Patrick H. Stakem, *Martian Railroad,* 2019, PRRB Publishing, ISBN-1794488243.

Patrick H. Stakem, *Exoplanets,* 2019, PRRB Publishing, ISBN-1731385056.

Patrick H. Stakem, *Exploiting the Moon,* 2019, PRRB Publishing, ISBN-1091057850.

Patrick H. Stakem, *RISC-V, an Open Source Solution for Space Flight Computers,* 2019, PRRB Publishing, ISBN-1796434388.

Patrick H. Stakem, *Arm in Space*, 2019, PRRB Publishing, ISBN-

<u>2019 Releases</u>

Extraterrestial Life

www.ingramcontent.com/pod-product-compliance
Lightning Source LLC
Chambersburg PA
CBHW030518220526
45464CB00006B/2852